EVERYONE EATS™

VEGETABLES

Jillian Powell

RAINTREE STECK-VAUGHN PUBLISHERS
The Steck-Vaughn Company
Austin, Texas

Titles in the Series

BREAD EGGS FISH FRUIT
MILK PASTA POTATOES
POULTRY RICE VEGETABLES

Published by Raintree Steck-Vaughn Publishers,
an imprint of Steck-Vaughn Company
Everyone Eats™ is a trademark of Steck-Vaughn Company

Library of Congress Cataloging-in-Publication Data
Powell, Jillian.
Vegetables / Jillian Powell.
p. cm.—(Everyone eats)
Includes bibliographical references and index.
Summary: Describes the qualities of various vegetables and methods of growing, harvesting, storing, and cooking them, with recipes.
ISBN 0-8172-4768-8
1. Vegetables—Juvenile literature.
2. Cookery (Vegetables)—Juvenile literature.
[1. Vegetables. 2. Cookery—Vegetables]
I. Title. II. Series: Powell, Jillian. Everyone eats.
TX401.P68 1997
641.6'5—dc21 96-29687

Printed in Italy. Bound in the United States.
1 2 3 4 5 6 7 8 9 0 01 00 99 98 97

Picture acknowledgments

Cephas 4, 7 (bottom), 8 (both), 11 (top), 15 (top), 16, 19 (bottom), 20, 22 (both), 23 (top), 25 (both); Chapel Studios 6, 10 (top), 11 (bottom), 17 (top), 18, 19 (top), 21 (top), 23 (both), 25 (top); Mary Evans 9 (left); Eye Ubiquitous 10 (bottom), 12, 13 (bottom), 14 (both), 15 (bottom), 21 (bottom), 24 (both); Imperial War Museum 9 (right); Life File contents page, 5 (bottom), 7 (top), 13 (top), 17 (bottom), 18 (right), 24 (top); Wayland Picture Library title page, 5 (top)

Contents

Varied Vegetables

◀ A colorful selection of vegetables of all shapes and sizes. Different vegetables are grown in different countries, and there are hundreds of types of vegetables.

A vegetable is a plant, or part of a plant, that is used for food. Vegetables can be all shapes and sizes, including tiny peas and beans, leafy spinach and cabbage, and knobbly roots and tubers. They can be many colors, from red peppers, orange carrots, and yellow corn to green peas, purple eggplant, and black beans.

Vegetables have been an important food since early peoples gathered wild plants to eat. They contain lots of vitamins, minerals, and fiber, which we need to keep us healthy and to fight illness and disease.

Vegetable juices have been used to dye textiles and other materials since prehistoric times. The Victorians used beets to make a hair rinse and to dye clothes.

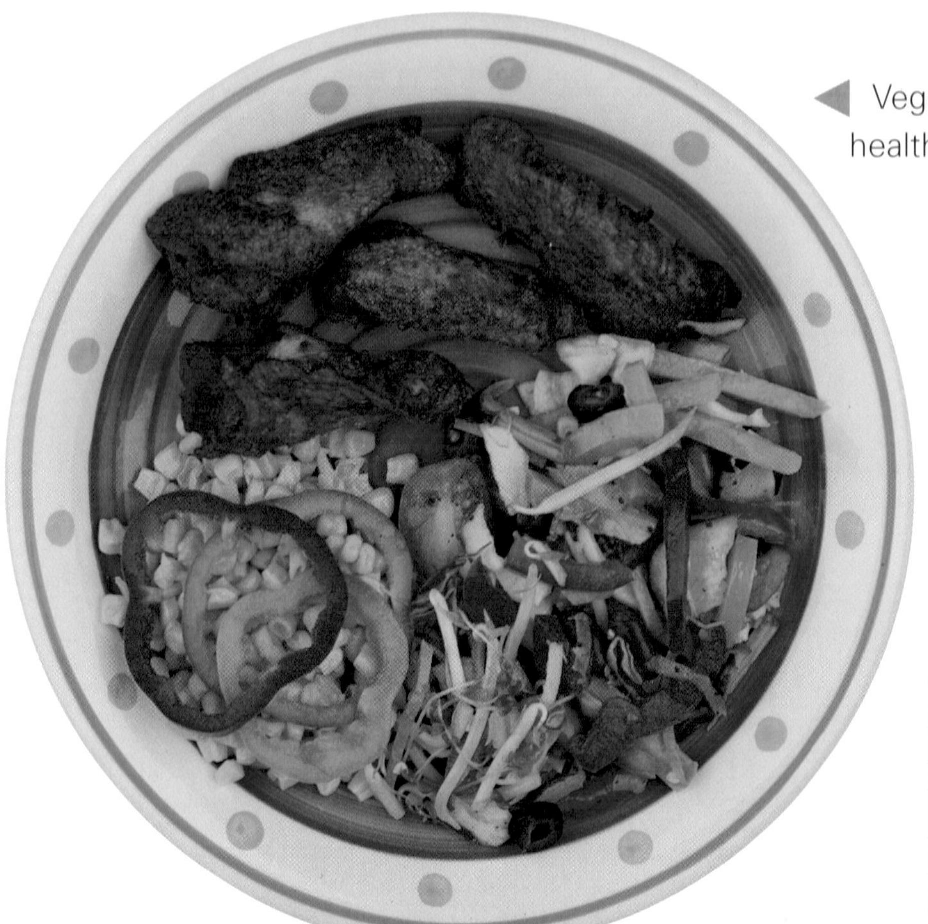

Vegetables are a colorful and healthful part of a balanced diet.

In the United States, each person eats about 220 pounds of fresh vegetables each year.

Vegetables are good to eat and have many varied flavors. They bring different colors, tastes, and textures to a meal. Some vegetables, like onions, chilies, and garlic, have a strong taste. Some can be eaten raw; others have to be cooked. Raw vegetables contain lots of nutrients.

We can serve vegetables with poultry, meat, fish, or vegetarian meals. They can also be used in dishes such as pies, casseroles, soups, burgers, and sausages, as well as pickles and chutneys, drinks, sauces, and salads.

Over 200 kinds of vegetables are eaten around the world. We can buy them fresh, frozen, canned, or dried. Fast transportation and cold storage make fresh vegetables from all over the world available all year round.

A farmer checking his homegrown leeks. People have been growing their own vegetables for thousands of years.

What Are Vegetables?

Vegetables are plants that we can eat. There are thousands of different plants growing all over the world. Plants start as seeds, which send roots down into the soil, and send shoots up to form stems that usually bear leaves and flowers.

Some foods that we call vegetables are really fruits, because they are the part of the plant that produces seeds. These include tomatoes, eggplant, cucumbers, pumpkins, and sweet peppers.

When we eat vegetables, we are eating part of the plant. We eat the roots of carrots, parsnips, radishes, and beets, the stems of celery and asparagus, and the leaves of lettuces, cabbages, spinach, and watercress.

Peas and beans are the seeds of plants that bear pods. The seeds can be dried to make pulses. Sometimes we eat the pods as well, as with green string beans, bush beans, sugar snaps, and snow peas.

◀ A vegetable is the edible part of a plant. These bean plants have leaves, flowers, a stem, and roots. The boy is picking beans to eat.

◀ Onions and garlic are bulbs—the rounded lumps from which some plants grow stems and roots. Fennel is another bulb vegetable.

Broccoli and cauliflower are flower heads. Brussels sprouts are buds, which grow on the stem of the plant.

▼ Root vegetables, including radish, parsnip, and beet

Plants like yams, Jerusalem artichokes, and potatoes produce food stores called tubers under the ground. The tubers are the parts we eat.

Fungi, including mushrooms and toadstools, grow from spores rather than from seeds. Some fungi are edible. Others can be poisonous. Never eat wild fungi or plants unless you are sure they are safe to eat.

Vegetables in the Past

▲ A field of corn. Corn, peas, and other pulses were among the earliest food crops.

Early peoples gathered vegetable plants from the wild and dug for roots and bulbs. In Neolithic times, from about 10,000 B.C., people learned to grow their own vegetables by collecting and planting the seeds.

In ancient China and parts of Europe, vegetables were salted or pickled, which prevented them from going bad, so they could be stored through the winter months. The Romans grew vegetables such as radishes, cabbage, carrots, and lettuce. They pickled vegetables in vinegar and honey and served salad and cooked vegetables with sauces.

In the Middle Ages, people grew onions and leafy and root vegetables, as well as edible herbs and flowers. Onions, pulses, and root vegetables were cooked in thick soups called pottages. Pease pottage, made with dried peas, was eaten during the winter months.

◀ A colorful herb and flower salad. In England from the fifteenth through the seventeenth centuries, people made flower salads containing primroses, violets, and nasturtiums.

More vegetables gradually became available, like corn, which the Spanish brought to Europe from South America, where it had grown for thousands of years. Farming methods improved during the eighteenth century, and by the end of the nineteenth century, the range of available vegetables had increased, through fast transportation, plant breeding, and the invention of canning and freezing methods.

In ancient Egypt, beans were believed to contain the souls of the dead. The ancient Greeks and Romans threw beans into a helmet as a way of casting votes at elections. White beans stood for a "yes" vote and colored beans for a "no" vote. Over 500 years ago, the Aztec people of Mexico used cacao (cocoa) beans as currency.

This nineteenth-century factory in Paris used machines to dry vegetables so they would keep longer.

During World War II (1939–1945) there were food shortages. People dug up lawns, flowerbeds, and roadside lanes to grow their own vegetables. In the United States "victory gardens" sprang up in backyards and vacant lots. Pamphlets were printed that taught people how to grow many vegetables in a small space.

During World War II, the British government printed posters like this, telling people to "Dig for Victory" by growing their own vegetables.

The Food in Vegetables

Vegetables are an important part of a healthful diet. They are rich in the vitamins and minerals we need to keep us healthy. Some of the vitamins in vegetables are antioxidants, substances that are thought to help us fight illness and disease. Vegetables, including sweet peppers, lettuce, and cabbage, are rich in the antioxidant vitamin C. This helps our body cells grow and repair themselves and fight illness. Vegetables provide up to 50 percent of the vitamin C and 25 percent of the vitamin A that are needed by the average person.

▲ Vitamin A is important for healthy growth, skin, and eyes. Dark green and orange vegetables, including spinach and carrots, are rich in vitamin A.

Cucumbers contain the minerals silicon and sulfur, which are said to help hair grow.

▶ Vegetables are best eaten soon after they are harvested, before they lose some of their vitamins. In many countries, vegetables are sold straight from the fields, as at this roadside stall in Sri Lanka.

◀ Vegetables help to balance the acids in high-protein foods like fish, poultry, and meat.

Vegetables are high in fiber, which helps us digest and pass food through our bodies. Fiber can also help reduce cholesterol in the blood. Cholesterol is a natural substance found in our bodies and in some foods. Too much cholesterol, which can be caused by eating lots of fatty and salty foods, can lead to heart disease. Vegetables are low in cholesterol and salt and contain almost no fat. They can contain up to 95 percent water and are low in calories.

A shortage of vitamin A can lead to night blindness—problems seeing clearly when there is little light. Carrots contain lots of vitamin A, in the form of carotene, which is why carrots are said to help us see in the dark.

▲ Vegetables that are rich in protein, such as soybeans and pulses like these, are important for vegetarians, who eat no meat.

Vegetables contain some protein, which we need to grow and repair our bodies, and some carbohydrate, which gives us energy. Soybeans contain twice as much protein as chicken, and three times as much as beef. For people who eat meat, vegetables can help to balance the acids made in the body by eating too much animal protein.

Farming Vegetables

Vegetables are grown on farms and in gardens, market gardens, and rented farmland. They are grown as food for people and farm animals. Sheep can be fed root crops like turnips in the winter, and cattle eat leafy kale and cabbage when grass stops growing in the autumn.

▼ As they grow, vegetables may need watering in dry weather. This Indonesian man is watering his crop of cabbages.

Some vegetables, like potatoes, can be grown in many parts of the world. Others, like yams and cassava, need a tropical climate; vegetables like peas, Brussels sprouts, and root vegetables need a temperate climate.

Like all plants, vegetables need sunlight, water, and air to grow. Plants take in sunlight and air through their leaves and water and food from the soil through their roots. These are turned into natural sugars by a process called photosynthesis. The sugars become starch to feed the plants, or cellulose, which makes up plant cell walls.

Land for growing vegetables is plowed in the autumn or winter to let air into the soil, to bury weeds, and to allow water to drain. This is usually done by a harrow, pulled by a tractor. The harrow breaks up the clods of earth so the soil is ready for planting. The farmer will usually spray the fields with animal manure or chemical fertilizers to feed the soil.

▲ On modern farms, seeds are sown by machines like this, called seed drills. Some farmers and gardeners, however, may still sow seeds by hand. Farmers usually sow seeds so crops will ripen at different times, to give several harvests.

Weeds can take food and water from vegetable crops, which can also be attacked by pests and diseases. Some farmers spray crops with chemicals to kill weeds, insects, and diseases. Organic farmers do not use chemical sprays because they believe the sprays harm people and the environment. Instead, they fight pests, disease, and weeds by using natural methods.

◀ In cooler climates, vegetables that could be killed by cold or frost, like this lettuce, may be grown under glass or in plastic tunnels.

The Vegetable Harvest

Vegetables must be harvested as soon as they are ready to eat. Some, like Brussels sprouts, and lettuce, have to be harvested by hand. Others, including peas, beans, and carrots, can be harvested by machines.

Leafy vegetables like cabbage are cut or pulled out of the ground.

Peas are usually harvested by machines, which cut the plants and then pick the pods from the stems and shell the peas.

▲ Root crops like these beets are harvested by machines, which cut off the leafy tops and lift the roots out of the ground. The leafy tops of vegetables like beets can be used for animal feed.

▶ A Vietnamese woman picking peas by hand

◀ A man in Spain winnowing kidney beans. Winnowing is a process of separating the beans from the pods and stems.

Pulse plants, such as lentils and some kinds of beans, are cut down and may be left to dry before they are threshed to separate the pods from the stems.

▼ On this farm in Wisconsin, corn is being stored in huge wire containers known as silos.

Once they are harvested, vegetables rapidly begin to lose water and vitamins. They must be quickly cleaned and packed ready for sale. Some are sold at the farm, but most are sent to markets and supermarkets for sale or to factories for processing. First, they are washed and trimmed, then they are sorted by size, weight, and quality, before being packed. Some vegetables may be stored on the farm in a clamp made of earth and straw or in refrigerated storage.

Storing Vegetables

There are many ways of processing vegetables to store them and to make cooking quick and easy. Traditional methods of preserving vegetables include bottling them in vinegar, wine, oil, or salt, drying them, or making them into pickles and chutneys. Chutneys are made by chopping vegetables and boiling them with spices, vinegar, and sugar.

▲ Vegetables being pickled in vinegar, an ancient way of preserving vegetables

In Germany, cabbage is pickled in salt to make sauerkraut. In Italy, olive oil is used for bottling vegetables, including sweet peppers and mushrooms. In China, many kinds of vegetables are pickled in rice wine.

▼ These corn cobs have been hung in the hot sun of Sri Lanka to dry.

Pulses like peas, beans, and lentils can be dried by hot air or in factory ovens. In hot countries like Africa, some vegetable crops are dried in the sun. Dried vegetables are smaller in size and weight than fresh vegetables and are easier to store and to transport.

Fresh vegetables can be frozen. Freezing destroys bacteria and locks in vitamins and minerals. Vegetables can be stored in a freezer for long periods, and are quick and easy to cook.

◀ The juice of some vegetables is extracted and sold in bottles or cartons as a nutritious drink. Some vegetable juice can be dried to a powder and used in ready-to-mix soups.

Vegetable Processing

Vegetables for freezing must be perfect, fresh, and ripe. They must be harvested and frozen within a few hours so that no nutrients are lost. At freezing plants, the vegetables are trimmed, washed, and peeled, then blanched—part-cooked in boiling water. Then they are cooled and frozen. Some vegetables are sliced or chopped first, and others are mixed together and sold in packs.

Peas should be frozen within two and a half hours of being picked. In the summer, they are harvested day and night by machines. Peas for freezing are harvested by machines called viners, which cut the plants and remove the peas from the pods. The peas are rushed from the fields to nearby freezing plants.

The first can of baked beans was made in the United States in 1895 by the firm H. J. Heinz.

◀ This sugarcane from Sri Lanka will be processed to make sugar. Sugar can also be made from the juice obtained by crushing beets.

◀ A selection of canned vegetables that are ready to heat and serve

Some vegetables are processed at canning factories. Canned vegetables can be stored for long periods and need only to be heated. Fresh vegetables are rushed from farms to canning factories. They are trimmed, washed, and peeled, then blanched and put into the cans. Liquid is added, which may contain salt, sugar, colors, and flavors. Then the can is sealed so that no air can get in. The cans are cooked at high temperatures to kill any bacteria, before they are cooled and labeled.

Vegetables can be processed to make vegetable products. Soybeans can be crushed to make soy flour, soy sauce, oil, margarine, and milk. They can also be made into chunks or granules, which are used instead of meat in vegetarian recipes. Bean meal, from crushed beans, can be used in animal feed.

▼ Tofu, which is made from processed soy beans, is popular in Chinese and Japanese cooking.

Canning was invented in the nineteenth century, as a result of experiments to store food for the French emperor Napoleon Bonaparte's armies.

Preparing and Cooking Vegetables

▲ A dish of raw vegetables, served with a dip. Raw vegetables are crunchy, colorful, and full of nutrients.

Salad vegetables like lettuce and celery can be eaten raw. Some crunchy or leafy vegetables like carrots and cabbage can also be eaten raw and may be chopped or grated and mixed with mayonnaise to make coleslaw.

All fresh vegetables should be washed and trimmed before they are eaten. Some, like onions, must be peeled first, but it is best to peel thinly because most of the nutrients they contain are just under the skin. Vegetables can be cooked in lots of ways, such as steaming, boiling, broiling, stir-frying, roasting, and stewing.

Steaming is one of the healthiest ways to cook vegetables. Steamed broccoli keeps about 80 percent of its vitamin C content, but boiled broccoli keeps only 33 percent. When vegetables are cooked in boiling water, vitamins and minerals are lost in the water. It is best to use as little water as possible and cook for only a short time.

▲ A pan of boiled potatoes. Some of the vitamins and minerals will be lost in the cooking liquid, so it is a good idea to use it to make soup or gravy.

Microwaving and stir-frying are both quick and easy ways of cooking vegetables and help to keep their vitamin content. The vegetables remain crisp and do not lose their color or shape.

▼ Vegetables like carrots, leeks, onions, and potatoes are often used to add flavor and texture to casseroles and stews.

Red kidney beans need thorough cooking because they contain a poisonous substance in their skin. Some pulses must be soaked in water for several hours before being cooked even longer in boiling water.

Root vegetables can be chopped and used in stews and casseroles, roasted in the oven in a little fat, or boiled and mashed. Vegetables like peppers, squash, and eggplant can be stuffed with meat or vegetables and baked.

Vegetable Dishes from Around the World

▲ Gazpacho and its ingredients

Many traditional dishes around the world use vegetables that grow in the area.

Gazpacho is a Spanish soup served cold and made with tomatoes, peppers, and cucumbers. Borscht is a Russian soup made with beets and other root vegetables and served with sour cream.

Dried beans were an important protein food for the Aztec and Inca peoples, and they are still widely used in Mexican and South American cooking. The Mexican dish of refried beans is made with kidney beans, onions, tomatoes, and chilies, mashed together and fried.

▶ A Mexican meal like this will usually include dishes made with corn, chilies, and beans.

Boston baked beans is a traditional American dish that was eaten by the first settlers. They cooked huge pots of beans in their bread ovens and flavored them with bacon, onions, and molasses.

In Vietnam and the Philippines, vegetables are usually cooked with a little pork or seafood to add flavor. The Koreans use soybeans and other vegetables to make spicy pickles.

The Indonesian dish gado-gado is a salad of cooked vegetables served with a hot, sweet-and-sour peanut sauce. In Indonesian markets, people sell baskets of vegetables flavored with coconut, chili, and lime.

▲ Cassoulet, a French dish made of goose or duck, pork, and white beans

In the Middle East, chickpeas are mixed with onions, herbs, and spices to make fried patties called falafel.

Yams are an important vegetable in Africa, the West Indies, and South America. They can be boiled, baked, broiled, or mashed to make fufu, a kind of dumpling served with soups and stews. African stews often contain okra, which is called "gumbo" in the Caribbean, where it is cooked with seafood and chicken.

▶ An Indian dish of curried okra. Okra is also known as "ladies fingers."

Festivals and Customs

▲ Vegetables are among the foods offered to the Hindu goddess Lakshmi at the festival of Diwali.

Vegetables are a traditional part of Thanksgiving and harvest festivals around the world. Since ancient times, they have been offered at shrines dedicated to religions like Buddhism or to local gods of the harvest. In Christian churches, vegetables are offered with bread and fruit to give thanks for the harvest.

In the South Pacific and many parts of Africa, yams are offered to the gods of the harvest during special festivals. On the Trobriand Islands in the Pacific Ocean, the yam harvest is celebrated with weeks of feasting and dancing. Women from each village compete for the title "best gardener" by decorating their bodies with paints and carrying their yams to be judged.

The leek became the national symbol of Wales in A.D. 640, when Saint David ordered his men to wear leeks in their hats during a battle with the Saxons. Some Welsh people still wear leeks on Saint David's Day (March 1).

◀ A colorful display of produce at a shrine in Tokyo, Japan, during the Harvest Festival

Vegetables are part of many traditional feasts. In the United States, pumpkin pie is eaten for the Thanksgiving feast. In Scotland, rutabagas (called "neeps") are served with haggis on Robbie Burns' night in January, which celebrates the Scottish poet's life. In the south of France, people make a sweet spinach tart to eat on Christmas Eve.

Irish colcannon is a potato, onion, and cabbage dish traditionally made for Halloween. It was once the custom to hide a wedding ring inside, and it was said that whoever found it would marry within the year.

In ancient times, mushrooms were thought of as "food of the gods." The pharaohs of ancient Egypt believed that mushrooms had magical powers, and the Chinese used them as medicine.

▼ In medieval times, many people thought that mushrooms were magical because they grow in the dark.

Vegetable Recipes for You to Try

Ratatouille

To serve four people, you will need:

1 onion, chopped
6 tablespoons oil
1 green pepper, chopped
1 red pepper, chopped
2 large eggplants, sliced
2 large zucchinis, sliced
4 large tomatoes or a 28 oz. can of plum tomatoes
salt and pepper
1 clove garlic, crushed
1 teaspoon mixed herbs

1 About an hour before you start cooking, sprinkle some salt on the sliced eggplant and zucchini. When you are ready to use them, wipe off the salty juices. This helps stop the vegetables from soaking up too much oil when they are cooked.

2 Ask an adult to help you to heat the oil in a saucepan. Add the onion and fry gently until it is soft.

3 Add the peppers and garlic, then the eggplant and zucchini. Stir them all together well. Add the salt, pepper, and herbs.

4 Cover the pan and cook for about 30 minutes. Check that the vegetables are not sticking to the pan, and stir from time to time.

5 Remove the lid of the pan and add the tomatoes. Stir well and cook for another 15 minutes without the lid.

Serve hot as a vegetable side dish or as a main dish served with baked potatoes.

Carrot cake

To make one cake, you will need:

Cake

$1\frac{1}{2}$ sticks (6 oz.) soft margarine
1 cup light brown sugar
3 medium eggs
$1\frac{1}{4}$ cup self-rising flour, sifted
2 level teaspoons baking powder
2 ripe bananas, mashed
$\frac{1}{4}$ lb. carrots, peeled and grated
$\frac{3}{4}$ cup walnut pieces or raisins

Icing

$3\frac{1}{2}$ tablespoons butter, softened
2 ounces cream cheese
$\frac{1}{2}$ cup confectioners' sugar
2–3 drops of vanilla

Note: Some people get sick if they eat nuts. Always tell people when you serve dishes containing nuts.

1 Ask an adult to help you set the oven to 250°F. Put the ingredients for the cake into a mixing bowl and beat together well for two minutes.

2 Grease a 9-inch square cake pan and pour in the cake mixture. Smooth out evenly.

3 Bake the cake for about one hour in the middle of the oven, until it feels firm and springy to the touch. Turn out the cake onto a wire rack and allow it to cool.

4 To make the icing, sift the confectioners' sugar into a bowl and add the butter, cream cheese, and vanilla. Mix together well, then spread over the cooled cake. Make a pattern on the icing with a fork.

Allow the icing to set before serving the cake.

Glossary

Aztecs Native Mexican peoples who lived more than 500 years ago.

bacteria Tiny living things that have only one cell. Some bacteria cause disease but others are useful.

calories Measurements of energy in food.

carbohydrates Starchy and sugary foods that give us energy.

casserole Food that is cooked in a covered dish in the oven.

cold storage A cool place where food may be stored to keep it fresh.

currency Money or other methods of buying and selling.

environment The landscape and the animals, plants, and people who live there.

fertilizers Substances used on crops to feed the plants.

fiber The part of food that helps us digest and pass the food through our bodies.

fungi A type of plant that grows from spores and does not make energy using photosynthesis.

haggis A Scottish dish made of ground or chopped meat, traditionally boiled in a sheep's stomach.

harrow A farm tool with teeth for breaking up plowed earth.

Incas Peoples who lived in Peru before the arrival of Spanish explorers in the sixteenth century.

market gardens Places where fruit and vegetables are grown to be sold.

medieval From the Middle Ages.

Middle Ages The period in history from the sixth to the fifteenth centuries.

minerals Substances found in some foods. We need certain minerals to keep us healthy.

molasses A kind of sugar syrup.

nutritious Containing food value.

organic farming A method of farming without using chemicals.

pharaohs Rulers of ancient Egypt.

prehistoric From ancient times, before written records were kept.

preserving Preparing for storage.

processing Preparing for storage or cooking.

protein The part of food that we need to build and repair our bodies.

pulses The edible seeds of certain crops, such as peas, beans, or lentils.

Saxons People from Europe who invaded England in the fifth and sixth centuries.

shrines Religious places.
spores Single cells from which some types of plants grow.
stir-frying Frying quickly in a lightly oiled frying pan or wok (large Chinese frying pan).
temperate Describing a region or climate that is never very hot or very cold.
threshed Beaten to separate seed or grain from a plant.
traditional Based on customs or beliefs handed down over time.
tropical Describing a region or climate with high temperatures and heavy rainfall.
tubers Lumps that grow underground on some plants.
vegetarian Eating no meat or animal products.
Victorians People living in England during the reign of Queen Victoria (1837–1901).
vitamins Substances found in some foods. We need vitamins to keep us healthy.

Books to read

Green, Harriet & Martin, Sue. *Sprouts*. Columbus, OH: Good Apple, 1981.

King, Elizabeth. *Chile Fever: A Celebration of Peppers*. New York: Dutton Children's Books, 1995.

Miller, Susanna. *Beans & Peas*. Food We Eat. Minneapolis, MN: Lerner Group, 1990.

Pulleyn, Micah & Bracken, Sarah. *Kids in the Kitchen: Delicious, Fun, & Healthy Recipes to Cook and Bake*. New York: Sterling Publishing, Inc., 1995.

Wolfe, Robert L. & Wolfe, Diane. *Vegetarian Cooking Around the World*. Easy Menu Ethnic Cookbooks. Minneapolis, MN: Lerner Group, 1992.

Index

Numbers in **bold** show subjects that appear in pictures.